Georges Seil, PhD

Understand and Design Energy Strategies

The Art of Modelling an
Energy
Strategy

Georges Seil, PhD
CEM®- CMVP®
CEO of BIZ-Consultant sàrl
L-8538 Hovelange
Grand Duchy of Luxembourg

ISBN: 9798634009476
Imprint: Independently published

Copyright © 2020 by Georges Seil
All rights reserved. No part of this publication may be reproduced, distributed, or transmitted in any form or by any means, including photocopying, recording, or other electronic or mechanical methods, without the prior written permission of the Author, except in the case of brief quotations embodied in critical reviews and certain other noncommercial uses permitted by copyright law. For permission requests, write to the author, addressed "Attention: Permissions Coordinator," at the address below.

BIZ-Consultant sàrl
Schmitzgaessel 2
L-8538 Hovelange
G.D. of Luxembourg

www.biz-consultant.net
georges.seil@biz-consultant.net

Preface

This book was written as an extension to the first book "Energy, Blockchain and Circular Economy," and targets all readers who are interested about the subject of developing energy strategies for all energy activity sectors. Practical use of energy canvas modelling and energy scorecards should provide the reader an easy understanding of this very often misunderstood subject.

Primary energy production, mainly based on fossil fuels like coal, gas and petrol or alternative energy production, mainly based on nuclear power are targeting local or/and regional markets. Renewable energy producers like solar, wind and biogas mainly target sales of renewable energy by feeding the production back to the grids. Part of this energy may be consumed at the production side as island power plant.

All producers have one common denominator: the customer. The energy producer with the best energy strategy will be the winner in the global energy contest.

However, it must be mentioned that the aim of this book is not to cover all scientific details, but it should provide evidence for the way it's done now, the way we'd suggest, and some alternative ways we might be able to do it.

Because of its interdisciplinarity and its easily kept explanatory texts, this book should be used as a key introduction into and a reference book for energy strategy designer, for energy policymakers, undergraduate and graduate studies, engineers, as well as casual readers.

The author would like to express special thanks to the Dean of Studies, of Rushmore University, Professor Alan Guinn for his coaching throughout the authors Doctoral Program and for the guidance in the writing of this book. Furthermore, also great thanks to Elizabeth Miller, the Rushmore University Editor who helped me structuring the content in a professional way.

Introduction Summary

Energy Strategy

An energy strategy is an action plan to manage the supply, procurement, cost, and efficiency of energy across all areas of a business. Energy strategy may be applied to multiple areas, like energy security strategy, energy efficiency strategy, energy production strategy, etc.

Energy Policy describes a set of rules for how a company or society should rely on the use and/or production of energy for complying with the long-term vision that is defined by the energy strategy.

By using existing and proven business management tools like the Business Model Canvas and the Balanced Scorecard, the book uses these tools for building a Techno-Economic Energy Model that allows developing a specific Energy Strategy related to energy production or energy efficiency in a distributed or a closed system.

This part of the book should also be used as a guide for developing innovative projects linked to EU programs like Horizon 2020, where project developers must rely on a given EU strategy and strictly follow a set of given objectives.

By using this Techno-Economic Energy Model, the project developer will have an increased chance of success in the project evaluation and selection process.

Tools that are described and used in this section include: Balanced Scorecard (page 49) and Business Model Canvas (page 26).

CONTENTS

Contents

- Abstract 8
- Introduction 10
- Techno-Economic Energy Model: Development Process 12
- Energy Model Canvas for Specific Types of Energy Business Models 13
  - EU Programs 13
  - Energy Generation for Global Energy Providers into Global Distribution Networks 17
  - Retail Energy Providers (REP) 17
  - Focused Renewable Energy Generation 17
  - Alternative Energy Generation 20
  - Energy Efficiency Projects for Industries 20
  - Energy Efficiency Projects for Buildings 22
  - Energy Certifications in Construction and Buildings 22
- Strategy Development Concepts in Business Environment 33
- Blue Ocean Strategy 34
- Porter's Generic Strategies 36
- EU Horizon 2020 programs 40
- Energy generation for global energy providers into global distribution networks 41
- Conclusion about comparing both types of strategies 47
- eBSC Concept Development 48
- EMS-2.3 strategy converted to eBSC objectives 53
  - Technology Perspective 54
  - People Perspective 55
  - Process Perspective 56
  - Production Perspective 58
  - Customer Perspective 59
  - Emissions/Environmental Perspective 60
  - Impact Perspective 61

- Economic Savings Perspective 62
- Development of the Key Performance Indicators 62
- Conclusion 65
- **Bibliography and References** 66
- **Author's short Biography** 67

E-Strategy Design and Monitoring *Adapting Business Strategy Tools to the Domain of Energy*

Abstract

This book aims to explore the development of a practical tool to be used for both a) the development of a new Energy Strategy model related to a specific private, public, or business environment; and b) to implement an existing or required Energy Strategy in a given new or existing project environment. In relying on existing findings and proven tools that have been used for many years in corporate business strategy design, the book stresses the design of an Energy Strategy and adapting these tools to the Energy Environment.

The final purpose of the research is building a Techno-Economic Energy Model that allows development of a specific Energy Strategy related to energy generation or energy efficiency in a distributed or a closed system. Techno-economic energy models provide a holistic approach towards the configuration and operation of energy systems. Our model will allow us to identify the optimal trade-off between energetic, economic, and environmental performances, linked to given or newly designed energy strategy models. To our knowledge, there is no model available that has been custom-tailored for energy project design or for analyzing existing energy systems on maximum alignment to given energy strategies; therefore, it is necessary to not only adapt existing techno-economic energy models, but also to develop new models based on existing tools that are available today in Business and Economic environments.

For this purpose, existing tools and concepts like Balanced Scorecard, Business Model Canvas, and Michael Porter's Competitive Strategy Techniques will form the basis of the present book.

The research identified new options in existing strategy tools that were analyzed and that need to be further developed and presented in next books.

This book explains how to make use of existing Business Strategy Development and Monitoring tools in building a Techno-Economic Energy Model related to energy generation, energy efficiency strategy development and optimization, and as a guideline through the development process of an EU project proposal. In a subsequent book, each individual tool will be analyzed regarding its structure and features for best adapting to a modern energy strategy development process.

This book should also be used as a guide for developing innovative projects linked to EU programs like Horizon 2020, where project developers must rely on a given strategy and strictly follow a set of given objectives. By using this Techno-Economic Energy Model, the project developer will have an increased chance for success in the project evaluation and selection process.

The author used all described tools in the context of his own international projects related to Business Strategy Development and has already optimized specific tools for best performance in a Business environment. The challenge for this research is adapting business-oriented to energy-oriented methodology. Related trainings were provided by the author in Luxembourg, Croatia, Serbia, Bulgaria, Romania, Morocco, and the Ukraine.

Introduction

According to the European Commission's definitions on Energy Strategy[1], we can read the following strategy statements:

STS-1= [Building the Energy Union](#)
The Energy Union will help to provide secure, affordable and clean energy for EU citizens and businesses.

STS-2= [Clean Energy for All Europeans](#)
A package of proposed new rules aimed at providing the necessary legal framework to facilitate the clean energy transition.

STS-3= [Governance of the Energy Union](#)
Proposed new rules on the Governance of the Energy Union will help to ensure its objectives are met.

STS-4= [2020 Energy Strategy](#)
The EU has set 20% targets for renewable energy, greenhouse gas reduction, and energy efficiency for 2020.

STS-5= [2030 Energy Strategy](#)
The 2030 Energy Strategy proposes targets for renewables, energy efficiency, and greenhouse gas reductions for the period between 2020 and 2030.

STS-6= [2050 Energy Strategy](#)
EU strategy for the transition to a competitive, secure, and sustainable energy system by 2050 and for reducing greenhouse gas emissions by at least 80%.

STS-7= [Energy Security Strategy](#)
The EU Energy Security Strategy aims to ensure a reliable supply of energy for EU countries.

STS-8= [Clean Energy for EU Islands](#)
The Clean Energy for EU Islands initiative provides a long-term framework to help islands generate their own sustainable, low-cost energy.

Each of the above strategy statements provides specific elements to be focused on and to be considered when writing a project proposal. Let's have a closer look at STS-1. STS-1 is rather a less tangible strategy than STS-4, so it needs a more detailed view about converting less tangible strategies into tangible results.

---

[1] https://ec.europa.eu/energy/en/topics/energy-strategy-and-energy-union/building-energy-union

STS-1 Building the Energy Union - is made up of five closely related and mutually reinforcing dimensions:

1. Security, solidarity, and trust: diversifying Europe's sources of energy and ensuring energy security through solidarity and cooperation between EU countries
2. A fully integrated internal energy market: enabling the free flow of energy through the EU through adequate infrastructure and without technical or regulatory barriers
3. Energy efficiency: improved energy efficiency will reduce dependence on energy imports, lower emissions, and drive jobs and growth
4. Decarbonizing the economy: the EU is committed to a quick ratification of the Paris Agreement and to retaining its
5. Leadership in renewable energy research, innovation, and competitiveness: supporting breakthroughs in low-carbon and clean energy technologies by prioritizing research and innovation to drive the energy transition and improve competitiveness.

If we consider each one of the above dimensions as an individual strategic theme according to the Balanced Scorecard set of definitions[2], then we have the option of using five strategy maps for expressing our initial strategic themes.

In our case, we will consider two different factors:
1. The EU model, with an initial strategy available
2. The competitive model, by deriving the strategy from the Energy Model Canvas

The EMC will build the link from an energy strategy environment to one or more possible technological concepts that could provide the best results to our targeted Techno-Economic Energy Model, as the answer of a project proposal based on STS-1.

The book limits the research on the development of Energy Strategy STS-1 and 7 Energy Business Models.

---

[2] http://www.balancedscorecard.org/BSC-Basics/About-the-Balanced-Scorecard

Techno-Economic Energy Model: Development Process

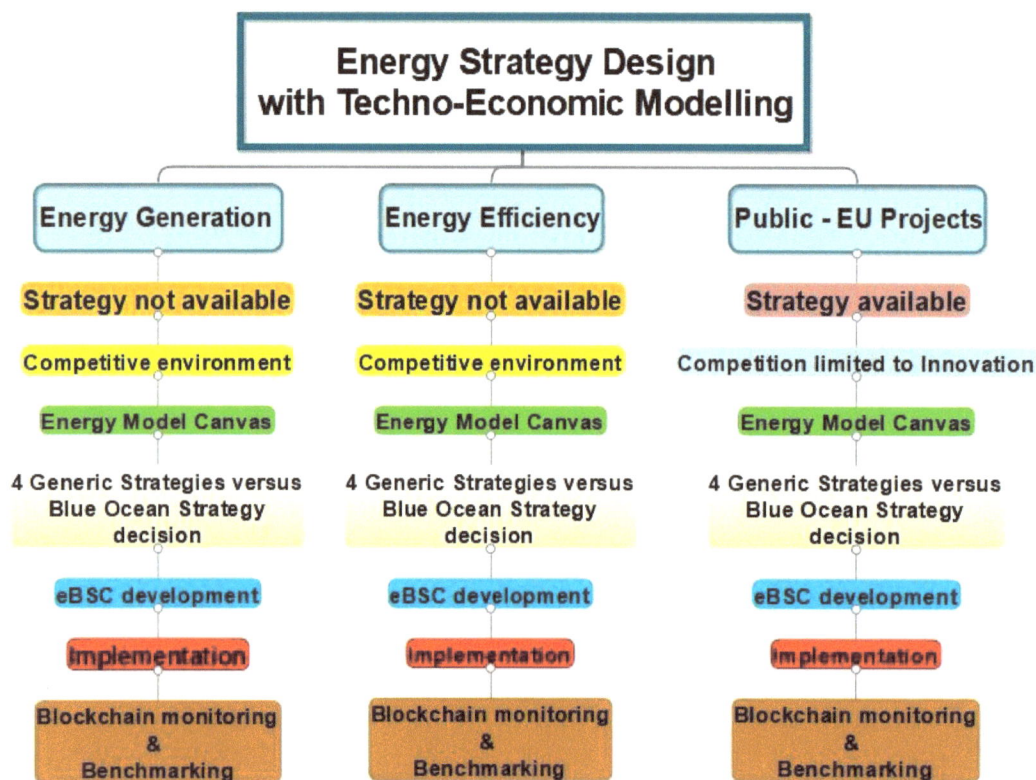

Fig.1: Structure of the Energy Strategy Design for Different Energy Models

Energy Model Canvas for Specific Types of Energy Business Models

We will differentiate between seven different Energy Business Models:
1. EU Horizon 2020 or related programs
2. Energy generation for global energy providers into global distribution networks
3. Retail Energy providers
4. Focused Renewable Energy generation
5. Alternative Energy generation
6. Energy Efficiency projects for Industries
7. Energy Efficiency projects for Buildings

EU Programs

Public projects, like the EU H2020 programs, follow a very defined strategy with specific objectives to be realized. In this case, all project developers and competitors must rely on the same basic strategy given by a specific EU program.

How can they gain competitive advantage?

Let's first use the definition of the competitive advantage[3]:

When two or more firms compete within the same market, one firm possesses a competitive advantage over its rivals when it earns (or has the potential to earn) a persistently higher rate of profit.

Competitive advantage, then, is the firm's ability to outperform rivals on the primary performance goal – profitability. But competitive advantage may not be revealed in higher profitability. A firm may trade current profit for investment in market share or technology.

In the case of EU projects, project developers compete against each other not at the level of profitability, but at the level of innovative concepts relying fully or partially on given objectives and strategies. EU programs

---

[3] Robert M. Grant, Contemporary Strategy Analysis, 2002, Blackwell Publishers

like Horizon 2020 require the following objectives to be completed at the level of the submitted proposal:

- Excellence
- Objectives of the proposal
- Relation to the Work Program
- Innovation
- Ambition
- Impact
- Dissemination and Exploitation of results.

In case of an EU project, two competitors proposing identical technology, identical cost structure, and identical project management strengths will be judged on the level of achievement and alignment to all above listed objectives. A major impact or a better dissemination process could tip the scales between both projects.

Coming back to our STS-1 strategy, we will now analyze the 5 Strategic Themes of STS-1 to build an energy concept. To better understand the STS-1-5 strategies, the author emphasizes that the strategic dimensions or themes each represent a separate project or a project that includes one or more of the strategic themes. For each theme we will annex a first possible strategy to be verified at a later stage in the strategy concept process.

In our case and for simplification purposes, we assume that each theme represents one single project.

STS-1 Building the Energy Union, is made up of five closely related and mutually reinforcing dimensions.

1. Security, solidarity, and trust: diversifying Europe's sources of energy and ensuring energy security through solidarity and cooperation between EU countries
    a. Key elements: Diversification, energy security, cooperation and solidarity
    b. Proposed strategy: to be analyzed in the strategy section
    c. Rationale: A diversification of energy sources and energy security through solidarity and cooperation between EU countries is targeted, which requires:
        1. Differentiation in competitive energy production and distribution networks

ii. Focus on superior quality; thus enabling secured operations
iii. Disseminate and communicate to include a wide consumer community

2. A fully integrated internal energy market: enabling the free flow of energy throughout the EU through adequate infrastructure and without technical or regulatory barriers[4]
   a. Key elements: integrated internal energy market, free energy flow, adequate infrastructures, no technical or regulatory barriers.
   b. Proposed strategy: to be analyzed in the strategy section
   c. Rationale: According to the EU Commission, a fully-integrated internal energy market requires (see footnote link) a "New energy market design," "empowering energy consumers, save money and have the choice for selecting providers," "consumer protection…," and "well being informed…;", demonstrating:
      i. Differentiation in competitive energy production and distribution networks
      ii. Focus on superior quality; thus enabling secured operations
      iii. Disseminate and communicate to include a wide consumer community
      iv. Leverage on energy efficiency

3. Energy efficiency: improved energy efficiency will reduce dependence on energy imports, lower emissions, and drive jobs and growth
   a. Key elements: improve energy efficiency, reduce dependence on energy imports, emissions, jobs creation
   b. Proposed strategy: to be analyzed in the strategy section
   c. Rationale: Use state-of-the-art technology at reasonable costs that allow best energy efficiency
      i. Differentiation in creating new energy efficiency models

---
[4] https://ec.europa.eu/commission/priorities/energy-union-and-climate/fully-integrated-internal-energy-market_en

ii. Focus on the ratio of savings/investment
iii. Disseminate and communicate new methodologies to all industry sectors
iv. Leverage on jobs creation and competitive advantage

4. Decarbonizing the economy: the EU is committed to a quick ratification of the Paris Agreement
   a. Key elements: reduce emissions, innovation towards climate-saving actions
   b. Proposed strategy: to be analyzed in the strategy section
   c. Rationale: Reducing emissions is not only dedicated to energy savings but will impact our entire living space. Isolated buildings have a low carbon output, but planting trees in an urban area is also an action towards lowering emissions; educating people towards adapting emission savings behavior fits to the same strategy.
      i. Differentiation in developing new mixed emission savings concepts
      ii. Focus on level of impact by considering entire populations, not limited to economic sectors
      iii. Disseminate permanently at all levels – people and economy
      iv. Leverage individual innovation actions

5. Leadership in renewable energy research, innovation and competitiveness: supporting breakthroughs in low-carbon and clean-energy technologies by prioritizing research and innovation to drive the energy transition and improve competitiveness
   a. Key elements: innovation and research towards renewable energies and decarbonization technologies
   b. Proposed strategy: to be analyzed in the strategy section
   c. Rationale: investing in innovation and research will be a key element for supporting all four prior strategic themes.
      i. Differentiation in the development of renewable and alternative energies, including energy storage and emission avoidance, as well as emissions reuse in industrial processes

ii. Disseminate to all related industry sectors for maximum impact
iii. Leverage R&D in storage, energy technologies, and emissions

Energy Generation for Global Energy Providers into Global Distribution Networks

Competitive advantage between energy operators in global distribution networks is much more focused on lowest price per kWh; therefore, lowest operating costs should be targeted for keeping operations profitable enough to allow investments and improvements of processes.

Cost factors at Global Energy Providers include high investments in technology, fuel costs, staff costs, distribution infrastructure, building infrastructure, and Research & Development.

Differentiating elements could include the provision of energy efficiency consulting and solving problems related to low Power Factor.

Retail Energy Providers (REP)

Competitive advantage between retail energy providers is focused on lowest price per kWh. As REPs only buy and sell energy from GEPs, they have no costs for production technology or R&D, and investments are generally limited to IT infrastructure, smart metering, and administration. Differentiating elements could include the provision of energy efficiency consulting, smart metering, and high-speed internet access at competitive rates.

Focused Renewable Energy Generation

The competitive advantage between focused renewable energy producers substantially varies amongst the technologies used for the production. We should consider both types of low carbon-emission technologies: alternative energies and renewable energies. Alternative energies are based on low carbon-emission technologies but indicate the supply of its primary energy resource is limited.

Examples of renewable sources of energy are solar, wind, and hydro. Also, Biogas is considered a renewable source of energy, even though its primary energy source is crops, grass, and food waste. These are substances that initially need to be produced prior to entering the process of energy generation.

Examples of alternative energy sources are natural gas turbines, nuclear power, natural gas co- and tri-generation, pyrolysis from plastics and rubber, waste processing, and fuel cells based on natural gas.

Competitive advantages for renewable and alternative energy production vary based on many parameters. Those parameters start from each country's grant and subsidizing system for specific renewable and/or alternative technologies and extend to the productivity and technology-based efficiency to the purchasing costs of the primary energy substances. In other words, a same technology could be profitable and competitive in one country and be unprofitable and non-competitive in another country.

Again, a simplified example should clarify this statement. This example uses Solar Photovoltaics in the Ukraine and in Luxembourg by two competitor EPCs [5]:

Ukraine – Solar PV
Feed-In: €180/MWh – during the length of the contract
Technology mainly from China
Subsidies for technology - €0.0

Electricity tariff for households: €40/MWh

Investment for 10kWp: €25,000

Luxembourg – Solar PV
Feed-In – in 2018: €259/MWh on 15 years
Technology from China and Europe
Subsidies for technology 10% of investment ≤€50,000

Electricity tariffs for households: €160/MWh

Investment for 10kWp: €23,000
Same as Ukraine less 10% subsidies

---

[5] EPC = Engineering Procurement Construction company

Assumptions: same basic investment costs as for Luxembourg; less subsidies advantage

Solar irradiation in Kiev: 1095 kWh/kWp/year

Solar irradiation in Luxembourg city: 1000 kWh/kWp/year

Payback: 10-11 years

Payback: 6-7 years

Table 1: Simplified example of different country feed-in and subsidizing schemes

Using both PV systems in the same way will make the project successful for Luxembourg, but for the Ukraine the project payback time will be too high and thus will not be profitable for the client.

What could be a differentiator between selling the project to the Ukraine and to Luxembourg?

- Solar irradiation is a climate factor and independent of the EPC
- The costs of the panels are the same in both countries
- Feed-in tariffs are fixed by the regulator/government
- Subsidies are a matter of the government
- Electricity tariffs are fixed by the regulators and energy operators

What are the variables?
- Investment financing: offering the customer reduced-interest financing
- Using solar modules with higher efficiency (Monocrystalline versus Polycrystalline; combined solar modules with liquid heating capability)
- Selling the customer, the shadowing advantage of the solar modules with the effect of cooling rooftops in summertime, thus reducing energy consumption for air-conditioning.
- Renting the customer's roof and paying the customer a renting fee for a limited period.
- To be continued

Alternative Energy Generation

The competitive advantage in alternative energy generation systems is mainly located in converting innovation to beyond-state-of-the-art technologies and projects. An example of converting waste plastics and used car tires to syngas and synthetic fuel by pyrolysis technology can show different approaches to the construction of the reactor, different technologies to be used for the heating concept, the distillation process, the feeder, etc.

Going beyond-state-of-the-art in pyrolysis could mean the development of a process for decomposing syngas from a mixture of many different hydrocarbons that can be used as a feedstock for producing hydrogen. Hydrogen has the advantage of being a primary energy resource with increasing consumption in the industries. Using syngas only for producing electric and thermal energy is currently considered a commodity process. The profitability of such a basic pyrolysis system is limited to the feed-in tariffs of electric energy and the processes where the thermal energy can be used with profit.

If we consider a focus on thermal energy usage, then we approach a new EU program area: the waste heat energy usage. This area is already filled with many innovative projects, and a minority of these EU program-related proposals have been awarded with EU project co-financing in 2017. Thus, there is still major potential for good project proposals available.

Energy Efficiency Projects for Industries

The first focus on energy efficiency projects for industries is on:

- Optimizing existing processes
- Keeping investments as low as possible
- Maximum payback periods should not exceed 3-5 years
- Equipment older than 10 years should be replaced by high-efficiency equipment

Competition in the sector of Energy Efficiency comes from the entire Energy Consulting family, both professionals and non-professionals. The competitive advantage for Energy Efficiency can only be obtained by offering the customer the best consulting know-how and competence profile. This is only possible if the consultant shows a strong engineering background, topped by at least one or more of the following certifications:

- A minimum certification on Energy Efficiency; the best option is the C.E.M. certification by the Association of Energy Engineers. This is an international certification, and certification courses are currently available in over 25 countries worldwide
- Optional: CMVP (Certified Measurement and Verification Professional)
- Optional: ISO 50001 Auditor

In addition to the certifications the energy consultant should demonstrate strong reference projects and be competitive on his/her daily tariff fees.

A strong competitive advantage is the number of references for EE projects with a total percentage on energy savings that the consultant can demonstrate to his client and the ratio between savings and € of investment. A total percentage of savings instead of a total amount in MWh/year or tons of emissions savings is justified in how the percentage demonstrates the high or low related performance improvement achieved by implementing the Energy Efficiency Measures on a given process. In other words, if the Energy Expert has achieved 50% savings on an installed 1000 kWh/year process, the total savings in kWh will be 500kWh/year. But for the same industry customer, its sister company operates a 10,000 kWh/year process, the savings for the sister company will be 5,000 kWh/year. The demonstration should explain the importance of the process improvement, generating additional potential savings on identical processes instead of just looking to an instant-picture view of one single savings opportunity.

Energy Efficiency Projects for Buildings

This market segment is mainly in the hands of architects and construction EPCs. In almost all 28 EU countries, buildings, energy efficiency and energy savings programs have been regulated over all building construction concepts. Energy specialists are in most cases certified by local agencies, and accordingly the certification is only recognized at a national level. Energy Experts are selected directly by the municipality for working out the public infrastructure Energy Efficiency program, awarded by the European Energy Award program.

The European Energy Award supports municipalities willing to contribute to sustainable energy policy and urban development through the rational use of energy and increased use of renewable energies. There are more than 1,400 municipalities participating today.

A well-known Energy Efficiency program in the UK is ESOS, the Energy Savings Opportunity Scheme (ESOS) is a mandatory energy assessment and energy-saving identification scheme for large undertakings (and their corporate groups). The scheme applies throughout the UK.

Competitive advantages in this sector can be achieved by offering the following values:

- CEM certification in addition to the local EE certification
- CMVP certification
- HVAC experience
- Construction engineering experience
- Reference projects
- For private projects: Price
- For public projects: relation with public entities

Energy Certifications in Construction and Buildings

Buildings have a huge impact on the health and well-being of people and the planet. Buildings use resources, generate waste, and are expensive to maintain and operate. Green building is about designing, building, and operating buildings to maximize occupant health and productivity, use fewer resources, reduce waste and negative environmental impacts, and reduce costs.

Starting with the planning of the object, it is primarily the architect who must ensure the criteria influencing the energy consumption by taking measures in terms of thermal insulation of the envelope, ventilation and heat recovery techniques, the building's geographical position, ideal shading and an effective and efficient system for the production of heat and / or cold. Thus, it is the architect who should first be trained and certified in the energy design of the building.

Since 1998, Leadership in Energy and Environmental Design (LEED) has been one of the most popular green building certification programs in the world. Developed by the US Green Building Council (USGBC), a non-profit organization, it includes a set of evaluation systems on four levels (Certified, Silver, Gold, Platinum) for the design, construction, operation, and maintenance of buildings, housing, and green neighborhoods, which aims to help building owners and operators become environmentally friendly and use resources efficiently.

In Europe, the "European Energy Award" was introduced in 2005 and is the basis of the energy certification of municipalities for buildings and public infrastructures. LEED procedures have been used basically for EEA certification.

In 2007, the Energy Performance Certificate (EPC) or Energy Passport for Buildings, was introduced in Europe as a result of the preliminary reflections and results of the EEA. In those days, the CPE was not obligatory in all the countries of Europe. The certification is classified according to the same principle as the electrical appliances, in particular class A for best performance and class I for lowest performance. There is no regulated expertise in Europe, but each country must define its own standards in terms of quality, knowledge, training, and competence of auditors / certifiers and energy experts.

The only country with a well-defined and strictly applied profile of experts and procedures in Europe is Great Britain, which has implemented the ESOS program (Energy Savings Opportunity Scheme) and which is strictly related to the consumption profile of buildings and technical installations.

ISO 50001: 2018 specifies requirements for the establishment, implementation, maintenance, and improvement of an energy management system in order to enable an organization to follow a systematic approach for the continuous improvement of energy performance, including energy efficiency, energy use, and consumption. As ISO 50001 is a tool and certification for energy management, it can be used to certify the constructor itself. In this case, the constructor, who is a major consumer of energy for consumers at construction sites such as cranes, concrete mixers, transport equipment, etc. can be certified on ISO 50001 in order to reduce its consumption and energy costs.

ISO 50001 is a tool for analyzing and managing energy for any energy consumption infrastructure. Thus, it could be proposed that the manager of a residential building uses the ISO process to manage in a sustainable way the energy consumption of the building and the behavior of the tenants and owners through training and distributed information.

In conclusion, there are a variety of different programs and certifications designed to reduce energy consumption in buildings, but in reality, only ISO 50001 is used to audit an energy system in a structured and well-defined way.

Now that we have analyzed in detail the attributes for each type of energy, we will proceed to the next step: The Energy Model Canvas.

Table 2: Business Model Canvas – EMC1 adapted to EU programs[6]

Key Partners	Excellence	Cross Cutting	Impact	Implementation
Countries	Clarity and pertinence of the objectives	Clear, realistic and measurable objectives	Any substantial Impact beside the Work Program	Quality and effectiveness of the Work Plan
Industry sectors	Soundness of the concept and credibility of the proposed methodology	Relation to Social Sciences and Humanities	Measurable, valued impacts expressed in GWh, Mtoe of emissions and resources	Extend to which resources assigned to Work Packages are in line with objectives/deliverables
Minimum number of Partners as set out in the call conditions	Demonstrate that proposal is BEYOND state of the art	Gender dimension	Enhance innovation capacity	Appropriateness of Management structures and procedures
Operational capacity for participating in the project	Appropriate consideration of interdisciplinary approaches	The strategic approach to international cooperation	Create new market opportunities	Risk Management
		science education	Strengthen competitiveness and growth	Innovation Management
		open access to scientific	Address climate change	Complementary of consortium participants
		ethics	Other for society	Appropriateness of allocation of tasks
		standardization	Quality of proposed measures to exploit and disseminate project results	(valid role and adequate resources)
		climate and	IPR Communicate project activities to different target audiences	
	Key Resources		**Innovation**	
	# Human Key Resources		Level of Research	
	Financial Key resources - budgets		Integrate existing technologies to novel applications	
			Activities close to Market	
			Demonstration	
			Pilots	
			Proof of concept	

Cost Structure	Revenue Streams
What are the most important costs inherent in our	For what value are our customers willing to pay?
Which Key Resources are most expensive?	For what do they currently pay?
Which Key Activities are most expensive?	How are they currently paying?
	How would they prefer to pay?
	How much does each Revenue Stream contribute to overall

---

[6] Osterwalder, 2008, Business Model Generation. www.businessmodelgeneration.com

Table 3: Energy Model Canvas (EMC2)

Key Partners	Key Activities	Value Propositions	Customer Relationship	Customer Segments
1. BiSHEL District Heating provider and Electricity supplier 2. ELBIS Electricity distribution network operator 3. City of Luxembourg The local authority involved in urban planning regulations and their implementations	1. Local renewable heat & electricity generation, distribution and supply 2. Optimizing the production, storage/retrieving & buying/selling of energy 3. Supplying variable energy tariffs which reflect actual energy costs during the day 4. Supplying customer energy use information e.g.: a) high utilization of the district heating grid b) comparison of energy consumption to baseline group and customers own consumption history	Providing customers with: 1. Easy access to affordable renewable heat and electricity 2. Smart Metering services to help customers reduce their energy costs and footprint 3. Green sustainable eco-friendly local neighborhood/community	1. Energy supply contracts which offer a discount for buying both heat and electricity which reflect time of day energy production 2. Agreements with the local authority to use the district heating network	1. Residential customers (single family households) 2. Small commercial energy customers (shops, offices, schools, etc)
	Key Resources 1. A Bio-fuel powered CHP plant & electricity & heat distribution network 2. An energy management tool to optimize storage/retrieving and buying/selling energy in real time 3. Professional staff with required skills in marketing, engineering and finances 4. Certified Energy Managers		**Channels** 1. Local media TV, newspapers and large public screens, etc 2. Energy bills 3. Company Website 4. Handheld devices and websites for energy feedback 5. Electricity and heat distribution networks 6. Smart Meters	
Cost Structure 1. Renewable electricity and heat production costs & backup 2. Electricity purchased from day ahead market 3. Operating costs for the heat and electricity distribution 4. Maintenance and staffing costs 5. External auditing costs			**Revenue Streams** 1. Company attracts new customers by "Green Branding" and 2. Reduced costs for energy production and increased profits	

In Energy Model Canvas-EMC1 Table 3, we find all pertinent criteria required for writing a successful Horizon 2020 proposal related to energy efficiency and energy generation projects.

In other words, if the proposal includes all the above listed criteria and attributes concisely, then the chance of winning should already be well above 50%. The EMC-1 works like a checklist for applying all that is included in this canvas.

However, this EMC-1 is different than all other EMCs related to the Energy Models from 2-7, because in the other Energy Models, the EMC will become a tool for developing a competitive Energy Strategy, versus EMC-1, where the strategy is already defined by the EU program and the EMC has the role of a checklist for preparing the project proposal.

In both EMCs we find nine different cells; each cell is named by a Key Word that will be part of our strategy model, utilized as a guide for developing a strategy. An EMC is developed by a team in a Workshop. The team is composed of a team leader and team members. The team leader will ask the open questions for filling each individual cell with an unrestricted number of answers. Once all cells are completed, the next exercise will be the determination of the key strategic elements:

- For each revenue stream, what do we have to consider for achieving this target by using the elements from the Canvas?
- On which revenue streams should we focus?
- Should we go for a Blue or Red Ocean strategy?
- Assessment of the SWOT and PEST: how are the elements from SWOT and PEST diverging from or aligned to our attempted business model?
- Completing or adjusting the EMC with the findings from SWOT and PEST
- Writing a few words about what and where the draft Business Model will focus by using the Key Words as a guideline

Before we start thinking about how to develop a strategy, we must have completed the first steps of the pyramid:
1. Defining the Mission
2. Defining the Values
3. Defining the Vision

After this exercise, we can now start developing our strategy model.

Fig. 2: The strategy development pyramid

As the first step for building the strategy model is setting up the Business or Energy Model Canvas, we will now proceed by doing this exercise with EMC – (Table 4):

We start by using the cell
- Revenue Streams
    - and statement "1. Company attracts new customers by "Green Branding" and in turn lowers production costs."

The question is, which of the Canvas elements should we use for setting up a strategy towards "….attracting new customers by Green Branding…." ?

We continue to the next and following blocks and refer to Table 4

Key Partners	Key Activities	Value Propositions	Customer Relationship	Customer Segments
1. DDSSHEL District Heating provider and Electricity supplier 2. FI DAIS Electricity distribution network operator 3. City of Łukawica The local authority involved in urban planning, regulations and their implementations	1. Local renewable heat & electricity generation, distribution and supply 2. Optimizing the production, storage/retrieving & buying/selling of energy 3. Supplying the market with energy data which includes the actual energy costs including the day use information for the: a) highly urbanized part of the district b) local prosumers of energy group and customers in the group and customers in the industry 4. Supplying customers with energy	Providing customers with: 1. Easy access to affordable renewable heat and electricity 2. Smart metering services to allow customers to reduce their energy costs and footprint 3. Green sustainable eco-friendly local energy hub to boost local community	1. Energy supply contracts which offer a discount for buying both heat and electricity, which reflect in total daily energy production costs 2. Agreements with the local authority to use the district heating network	1. Residential customers (single family households) 2. Small commercial customers (shops, offices, schools etc.)
	Key Resources		Channels	
	1. A biofuel-powered CHP plant & electricity & heat distribution network 2. An energy management tool to optimize storage/retrieving and buying/selling of energy in real time 3. Professional staff with required skills in managing, engineering, finances 4. Certified Energy Manager		1. Local call center / TV, newspapers and the public (local events etc) 2. Energy bills 3. Company Website 4. Handheld devices and websites for energy feedback 5. Electricity and heat distribution networks 6. Smart meters	

Cost Structure	Revenue Streams
1. Renewable electricity and heat production costs 2. Electricity purchase based on daily market 3. Operating costs of the heat and electricity distribution networks 4. Maintenance and staffing costs 5. External auditing costs	1. Compensation from utilizing customers by 1) Green building hybrid inverter power production costs 2. Reduced costs for energy production and increased profits of reselling energy as a result of optimizing the energy production/storage/retrieving/distributing/selling of energy.

Table 4 Energy Model Canvas

- Key partners
  - ELDIS as the electricity distribution network operator
- Key Activities
  - Local renewable heat & electricity generation, distribution, and supply
- Value Proposition
  - Easy access to affordable renewable heat and electricity
  - Smart Metering services to help customers reduce their energy costs and footprint
- Customer Relationship
  - Energy supply contracts which offer a discount for buying both heat and electricity which reflects time of day energy production costs
- Customer Segments
  - Residential customers (single-family households)
- Channels
  - Local media TV, newspapers, large public screens, etc
  - Energy bills
  - Company website
  - Handheld devices and websites for energy feedback
  - Smart Meters
- Key Resources
  - A biofuel-powered CHP plant & electricity & heat distribution network
  - An energy management tool to optimize storage/retrieving and buying/selling energy in real time
  - Professional staff with required skills in marketing, engineering, and finances
  - Certified Energy Managers
- Costs Structure
  - Renewable electricity and heat production costs & backup non-renewable heat production costs
  - Maintenance and staffing costs
  - External auditing costs

Now, having all elements together will allow us to build a strategy and we can write our strategic statement as follows:

- By focusing on the development and growth of the green energy market, we will partner with ELDIS as the distribution network operator. Our key activities will be the generation of renewable heat and electricity, and we will be the one-stop shopping point for generation, distribution, and billing. Our offer is based on our easy access to affordable renewable heat and electricity and providing energy supply contracts with discount for buying both heat and electricity. We will mainly focus on residential customers prior to making the next step to professional sectors. The communication channels to be used are local media TV, newspapers, and large public screens, as well as messages on energy bills, updates on our website, customized messaging to handheld devices, and the use of Smart Metering. Our Key Resources are the power generation by our owned CHP plant and our owned heat distribution network beside ELDIS with the electricity distribution network. Our energy management tool will be deployed by our specialized staff and our Certified Energy Managers will provide energy efficiency advice to our private clients. Our cost structure will include the production and maintenance costs, as well as our internal administration and technical staff and in addition to external auditing costs required by new project developments.

Once this exercise is completed, we will proceed to the next step, which is evaluating the strategy statement related to a fitting Business Model.

For this exercise, we will analyze Porter's generic strategy model and the Blue Ocean[7] strategy philosophy.

---

[7] https://www.blueoceanstrategy.com/about-the-authors/

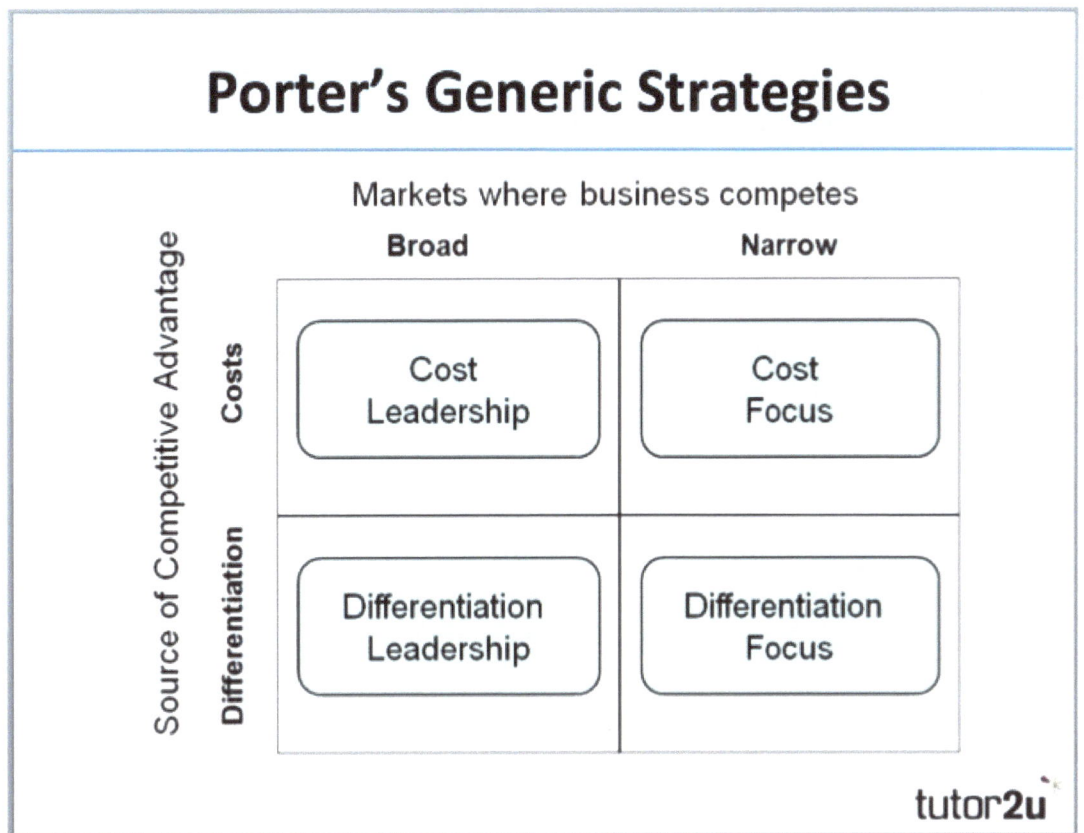

Fig 3: M. Porter, Generic Strategies
[Source: TUTOR2U – study notes
https://www.tutor2u.net/business/reference/porters-generic-strategies-for-competitive-advantage]

Blue Ocean Strategy

Red Ocean Strategy	Blue Ocean Strategy
Compete in existing Markets	Create new Markets
Knock out the competition	Make competition irrelevant
Valorize and exploit existing demand	Create new demand by innovation and differentiation
Make the value-cost trade-off (either have a value or low costs, not both)	Break the value-cost trade-off (have both high value and low costs)
Align the whole system of a firm's activities with its strategic choice of differentiation OR low cost	Align the whole system of a firm's activities towards differentiation AND low cost

Fig 4: Red versus Blue Ocean Strategy[8]

Are M. Porter's Generic Strategies still valid for an energy environment?

Kim and Mauborgne[9] argue that the strategy of Porter is not the right approach for a profitable growth in many markets. When the appeal is unfavorable, a structuralist approach is not an option. This often occurs in sectors characterized by oversupply and homogeneity of value propositions. In such situations, a company should adopt a reconstructionist approach and create a strategy to redefine the sector boundaries.

According to Davide Dall'Agata [Critique of the book "Blue Ocean Strategy," Research Gate, October 2015], Porter's vision of

---

[8] [Source: Renée Mauborgne and W. Chan Kim, 2015, Harvard Business Review Press]
[9] W. Chan Kim and Renée Mauborgne, Blue Ocean Strategy, Harvard Business School Press 2005

the strategy is assimilated to the structuralist economic theory, based on the paradigm of structure-conduct-performance.

Since the Industrial Age, firms have always struggled for the acquisition of advantages against the competition. The authors define this as the "red ocean" of competition: an ocean full of rival companies in an increasingly tight market at lowering profitability. Conversely, if companies could operate without any competition, the scenario would be quite different. In blue oceans, companies must create their own innovative models which will allow them to work in temporary uncontested markets, where growth is guaranteed.

By going back to the subject of Energy, we can observe both theories to be valid for describing an Energy strategy. In the case of the global players of energy generation and distribution, we are in a full competitive commodities market, the worst and most unprofitable market ever. This is the Red Ocean Market. In this case, Porter's Generic Strategies concept makes sense.

But when considering EU programs for project development, the competitive element is innovation instead of profitability. In this case, Porter's strategic concept needs to be adjusted or made more relevant to be replaced by the Blue Ocean Strategy. Blue Ocean Strategy is focused to step out of existing concepts and thrive on developing new spaces.

We will go through each one of Porter's Generic Strategies statements for **the seven types of energy operating segments** and testing them by its mirror statements on energy. After the Porter exercise, we will do the same exercise with the Blue Ocean Strategy.

Porter's Generic Strategies

Cost Leadership

Business	Energy
With this strategy, the objective is to become the lowest-cost producer in the industry	Use State-of-the-Art technology at low costs by targeting best possible efficiency
High levels of productivity	High level of productivity on generation and efficiency
High capacity utilization	High capacity utilization
Use of bargaining power to negotiate the lowest prices for production inputs	Use of bargaining power to negotiate the lowest prices for production inputs
Lean production methods	Lean production methods for generation and energy efficiency projects building
Effective use of technology in the production process	Effective use of technology in the production process

Cost Focus

Business	Energy
Business seeks a lower-cost advantage in just one or a small number of market segments	Focus on 20% of technology, impacting 80% on energy efficiency. For the production, use available technology at lowest costs by targeting highest energy efficiency
Product will be basic but acceptable by all customers	Product will be basic but acceptable by all projects

Differentiation Focus

Business	Energy
Differentiate within just one or a small number of target market segments	Differentiate at the level of the technology for a specific user group or market
Provide products that are clearly different from competitors who may be targeting a broader group of customers	Avoid using practiced technologies from competitive projects and focus on developing new technologies or technologies at Technology readiness level 7 (TRL)

Business	Energy
Differentiation focus is the classic niche marketing strategy	Use niche markets in efficiency by focusing on process optimization prior changing technology. In Energy Production focus on undeployed technology markets like plastics and rubber pyrolysis

Differentiation Leadership

Business	Energy
Business targets much larger markets and aims to achieve competitive advantage through differentiation across the whole of an industry	Differentiate at the level of the technology or application for a specific user group or a market with high potential of innovation that provides highest impacts. Energy Efficiency targets optimization processes across the whole of an industry
Superior product quality (features, benefits, durability, reliability)	E-production: focus on superior product quality and efficiency based on the Life Cycle Analysis

Business	Energy
Branding (strong customer recognition & desire; brand loyalty)	Branding (strong customer recognition & desire; brand loyalty) – the consultancy for EE and the EPC's (Engineering, Procurement and Construction companies) for project construction and design
Industry-wide distribution across all major channels (i.e. the product or brand is an essential item to be stocked by retailers)	For energy production = focus on the distribution on micro grids and feeding into global E-Networks. For Energy Efficiency = focus on the policy-makers for improving EE in industrial processes
Consistent promotional support – often dominated by advertising, sponsorship, etc	For both EE and Energy production project development, advertising is rarely adopted but sponsorship should be focused on specialized industries related to the project subject

The above analogy between Business and Energy demonstrates that Porter's Generic Strategies can be very helpful and can be used for expressing our energy strategy as a "Generic Energy Strategy." However, when comparing the above energy project/segments to both Porter's and the Blue Ocean strategy, we will see a first trend and a following situation:

EU Horizon 2020 programs

Porter
Cost Leadership: Not relevant
Cost Focus: Not relevant
Differentiation Focus: Highly relevant
Differentiation Leadership: Highly relevant

Blue Ocean Strategy

Red Ocean Strategy	Relevance	Blue Ocean Strategy	Relevance
Compete in existing Markets	NR	Create new Markets	High
Knock out the competition	Low	Make competition irrelevant	High
Valorize and exploit existing demand	Low	Create new demand by innovation and differentiation	High
Make the value-cost trade-off (either have a value or low costs, not both)	Low	Break the value-cost trade-off (have both high value and low costs)	High
Align the whole system of a firm's activities with its strategic choice of differentiation *OR* low cost	Medium	Align the whole system of a firm's activities towards differentiation *AND* low cost	High

Porter

Cost Leadership: Highly relevant
Cost Focus: Highly relevant
Differentiation Focus: Highly relevant but limited
Differentiation Leadership: Highly relevant but limited

Blue Ocean Strategy

Red Ocean Strategy	Relevance	Blue Ocean Strategy	Relevance
Compete in existing Markets	High	Create new Markets	High New option
Knock out the competition	High	Make competition irrelevant	Low
Valorize and exploit existing demand	High	Create new demand by innovation and differentiation	Medium
Make the value-cost trade-off (either have a value or low costs, not both)	High	Break the value-cost trade-off (have both high value and low costs)	Low
Align the whole system of a firm's activities with its strategic choice of differentiation _OR_ low cost	High	Align the whole system of a firm's activities towards differentiation _AND_ low cost	Low

Retail Energy providers

Porter
Cost Leadership: Highly relevant
Cost Focus: Highly relevant
Differentiation Focus: Highly relevant
Differentiation Leadership: Highly relevant

Blue Ocean strategy

Red Ocean Strategy	Relevance	Blue Ocean Strategy	Relevance
Compete in existing Markets	High	Create new Markets	High
Knock out the competition	High	Make competition irrelevant	High
Valorize and exploit existing demand	High	Create new demand by innovation and differentiation	Low for core business
Make the value-cost trade-off (either have a value or low costs, not both)	High	Break the value-cost trade-off (have both high value and low costs)	Medium
Align the whole system of a firm's activities with its strategic choice of differentiation OR low cost	High	Align the whole system of a firm's activities towards differentiation AND low cost	High

Focused Renewable Energy Generation

Porter
Cost Leadership: Relevant
Cost Focus: Relevant
Differentiation Focus: Highly relevant
Differentiation Leadership: Highly relevant

Blue Ocean Strategy

Red Ocean Strategy	Relevance	Blue Ocean Strategy	Relevance
Compete in existing Markets	High	Create new Markets	High
Knock out the competition	Low	Make competition irrelevant	Medium
Valorize and exploit existing demand	Medium	Create new demand by innovation and differentiation	High
Make the value-cost trade-off (either have a value or low costs, not both)	Medium	Break the value-cost trade-off (have both high value and low costs)	High
Align the whole system of a firm's activities with its strategic choice of differentiation *OR* low cost	Medium	Align the whole system of a firm's activities towards differentiation *AND* low cost	High

Alternative Energy Generation

Porter
Cost Leadership: Not relevant
Cost Focus: Not relevant
Differentiation Focus: Highly relevant
Differentiation Leadership: Highly relevant

Blue Ocean Strategy

Red Ocean Strategy	Relevance	Blue Ocean Strategy	Relevance
Compete in existing Markets	Low	Create new Markets	High
Knock out the competition	Low	Make competition irrelevant	High
Valorize and exploit existing demand	Low	Create new demand by innovation and differentiation	High
Make the value-cost trade-off (either have a value or low costs, not both)	Low	Break the value-cost trade-off (have both high value and low costs)	High
Align the whole system of a firm's activities with its strategic choice of differentiation *OR* low cost	Low	Align the whole system of a firm's activities towards differentiation *AND* low cost	High

Energy Efficiency Projects for Industries

Porter
Cost Leadership: Highly relevant
Cost Focus: Highly relevant
Differentiation Focus: Highly relevant
Differentiation Leadership: Highly relevant

Blue Ocean Strategy

Red Ocean Strategy	Relevance	Blue Ocean Strategy	Relevance
Compete in existing Markets	High	Create new Markets	Low
Knock out the competition	High	Make competition irrelevant	High
Valorize and exploit existing demand	High	Create new demand by innovation and differentiation	High
Make the value-cost trade-off (either have a value or low costs, not both)	Low	Break the value-cost trade-off (have both high value and low costs)	High
Align the whole system of a firm's activities with its strategic choice of differentiation *OR* low cost	Low	Align the whole system of a firm's activities towards differentiation *AND* low cost	High

Energy Efficiency Projects for Buildings

Porter

Cost Leadership: Highly relevant
Cost Focus: Highly relevant
Differentiation Focus: Highly relevant
Differentiation Leadership: Highly relevant

Blue Ocean strategy

Red Ocean Strategy	Relevance	Blue Ocean Strategy	Relevance
Compete in existing Markets	High	Create new Markets	High — For problem solving expertise
Knock out the competition	High	Make competition irrelevant	Low
Valorize and exploit existing demand	High	Create new demand by innovation and differentiation	Low
Make the value-cost trade-off (either have a value or low costs, not both)	High	Break the value-cost trade-off (have both high value and low costs)	Medium
Align the whole system of a firm's activities with its strategic choice of differentiation OR low cost	High	Align the whole system of a firm's activities towards differentiation AND low cost	Medium

Conclusion about comparing both types of strategies

Blue Ocean and competitive strategies overlap, and managers do not face a distinct either/or decision between each strategy.

In value innovation (Blue Ocean), the competitive strategy framework could suggest the contrary. For this reason, you can enter a "blue ocean" of uncontested markets with an opportunity-maximizing and risk-minimizing strategy. Hence, to compete successfully, a value innovation is necessary to escape the "red ocean" trap. As a result, you will drive costs down while simultaneously driving value up for buyers by reconstructing industry boundaries. In other words, the innovation framework will allow you to differentiate while simultaneously lowering your cost.

Using Porter or Blue Ocean depends on the type of business and the ability to innovate or compete.

eBSC Concept Development

The Balanced Scorecard (BSC) is a multidimensional approach to measure performance that incorporates both financial and non-financial factors. With the Balanced Scorecard, organizations can easily see the cause-effect relationship between actions and outcomes, whether it's the effect a new Big Data analytical tool will have on financial performance or the impact a new process improvement will have on accelerating product development. No other management tool illustrates these interconnections with such clarity.

We know from Kaplan and Norton[10] that four perspectives are considered as the basis for a Business- oriented BSC system:

- Finance
- Customer
- Process
- People

The list should be read from bottom to the top, i.e. an economic/business system does not work without people, people rely on processes, processes maximize performance with the effect on cost reduction, and maximizing customer satisfaction and happy customers satisfies finances.

Analyzing the five principles of the Balanced Scorecard method shows the importance of deviating from a classical finance-only approach to a structured-leader approach:

1. Translate the strategy into operational terms
2. Align the organization to the strategy
3. Make the strategy everyone's daily job
4. Make strategy a continual process
5. Mobilize change through executive leadership

---

[10] **[The Balanced Scorecard, Robert S. Kaplan and David P. Norton, 1996]**

The company's mission statement explains why it exists. The vision states where it wants to be in the future. The strategy is the tool that defines the roles to execute the mission and to move towards the vision goals, implicating the organization as a whole and evolving over the time to meet the changing conditions posed by the real world.

The essence of strategy is to define, implement, and manage unique value propositions in order to differentiate a company from the competitors.
Splitting the strategy into strategic themes allows strategy segmentation. Segmentation increases with the number of increasing measurable elements, the quality of the feedback, and thus the accuracy of the system.

The common element for all strategies is to create shareholder value.

The value proposition describes the unique mix of product, price, service, relationship, and image that the provider offers its customer.
The development of this layer of the strategy map is essential to understanding the type of customer and their expectations, and may be summarized as follows:

1. Product leadership: reaching beyond the customers' dreams
2. Customer intimacy: knowing the people it sells to
3. Operational excellence: optimal combination of quality, price, and ease of purchase

The successful company excels in one of these three values and sets prioritization targets for the other two.
The value proposition enables the company to define its targeted customers.
The value chain describes the activity elements of the internal processes of a company.

Four Perspectives of Balanced Score Card

		Measures
Financial	Is the company achieving its financial goal?	Operating income Return on assets Sales growth Cash flow from operations Reduction in Administration expenses
Customer	Is the company meeting customer expectations?	Customer satisfaction Customer retention New customer acquisition Market share, On time delivery
Internal process	Is the company improving internal control process?	Defect rate Number of suppliers Material turnover Percent of practical capacity
Innovation	Is the company improving its ability to innovate?	Amount spent on employee training Employee satisfaction, retention Number of new products, number of patents, sales of new products as a % of total sales

Fig 5= Balanced Scorecard Perspectives for Business Environment[11]

By replacing "business" with "energy," we must use eight energy-related perspectives:

- Economic savings
- Impact
- Emissions / Environment
- Customer
- Production
- Process
- People
- Technology

Again, we start with the last item in the list:

- Technology: No energy concept without technology
- People operate the technology and thus will have to be trained on best usage and conformity
- People rely on repeating processes for maximizing the control on energy output and input

---
[11] source: Slideshare by LinkedIn

- Optimal aligned processes increase energy efficiency to best results and maximize production
- Maximizing production at highest efficiency is satisfying the customers and reducing emissions at minimum
- Satisfied customers use low emission energy services and become loyal customers
- Reduced emissions have an immediate environment impact
- Less environmental impact because of reduced emissions will have an immediate positive effect on economical savings throughout the entire energy production and consumption value chain

We will now use the Energy model EMC-Table 4 from page 31 as our Energy Business Model and compiling the Strategic Statement to eBSC management structure:

- By focusing on the development and growth of the green energy market, we will partner with ELDIS as the distribution network operator. Our key activities will be the generation of renewable heat and electricity, and we will be the one-stop shopping point for generation, distribution and billing. Our offer is based on our easy access to affordable renewable heat and electricity and providing energy supply contracts with discounts for buying both heat and electricity. We will mainly focus on residential customers prior to making the next step to professional sectors. The communication channels to be used are local media TV, newspaper, and large public screens, as well as messages on energy bills, updates on our website, customized messaging to handheld devices, and the use of Smart Metering. Our Key Resources are the power generation by our owned CHP plant, and our owned heat distribution network beside ELDIS with the electricity distribution network. Our energy management tool will be deployed by our specialized staff, and our Certified Energy Managers will provide energy efficiency advice to our private clients. Our cost structure will include production and maintenance costs, as well as our internal administration and technical staff, in addition to external auditing costs required by new project developments.

Balanced Scorecard Project

Strategic Map for Strategic Theme: Developing the Green Energy Market

Measures	Perspective		
ROI, EVA, Revenue, Earnings, Capital, Cash Flow	Economic savings	Research and Development is financed by the company's profit. R&D will boost new developments in the area of energy storage, energy management and grid optimization	Focus should not be set to maximizing profit, but to maximizing the value for the customer and allow a sustainable growth and profitability of the company. Part of the profit should be used for sponsoring ecological or/and energy related events and organizations showing a big prospect on development and dissemination of best practices
Positive Negative savings emissions	Impact	Less environmental impact as a result of reduced emissions will have an immediate positive effect on economical savings throughout the entire energy production and consumption value chain.	
Nuclear, Coal, Petrol, Electricity, Gas, Emission neutral, Green Certif-	Emissions	Reduced emissions have an immediate positive environment impact	Saving emissions will contribute to green certicates payments
Quality, Service, Pricing, Time, Image, Relations	Customer	Maximizing production at highest efficiency is lowering the costs, satisfying the customers, reducing emissions at minimum and keeping customers loyal	
Quality, Grid capacity, Metering, Capacity use, Efficiency, Storage	Production	Maximizing production at highest efficiency is lowering the costs and satisfying the customers	Production should be diversified to the use of independent sources of energy for reducing the risk of failure. Storage of energy is highly contributing to maximizing the production effiency
Production, Delivery, Marketing, Operations, Quality Control, Service Dept, R&D	Process	Optimally trained and experienced staff will help improving and best executing internal processes	Optimal aligned processes increase energy efficiency to best results and maximize production
Employee, HR Capital, Knowledge, Technology, Best Practices, Intangibles	People	An optimal staff training plan should be aligned to high level technical, operational, financial, Marketing and HR competence and performance increasing requirements	People operate the technology and thus will have to be trained on best usage and conformity. People rely on repeating processes for maximizing the control on energy output and input.
Innovation, Evolutive, Primary energy, Storage, Alternative, Renewable	Technology	Using an evolutive and innovative technology results in using low primary energy from food and agricultural residues	

Fig 6 Strategy Map of an Energy Project[12]

---

[12] Source: Georges Seil, BIZ-Consultant sàrl

EMS-2.3 strategy converted to eBSC objectives

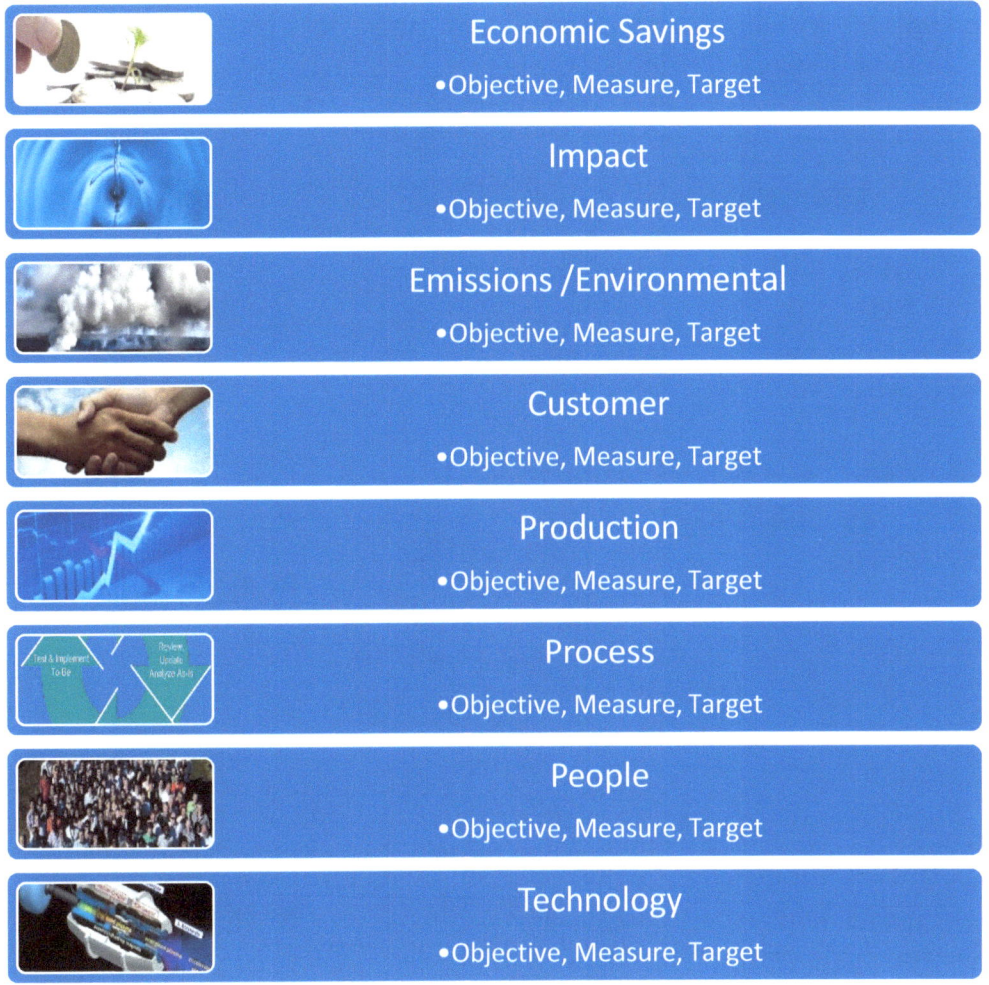

Fig. 7: The 8 Energy Perspectives[13]

The sequence of the eight perspectives follows the Balanced Scorecard Cause and Effect logic. Each perspective includes the attributes
- Objective
- Measure
- Target

In the following tables, we show how the EMC strategy is now developing objectives and extracting targets and measures.

The targets in this table refer to actions that must be completed to realize the related objectives.

---

[13] Source: Georges Seil, BIZ-Consultant sàrl

Measurements of completion or execution (one-time or permanent) will be handled by the Key Performance Indicators (KPI), which are explained in the section following the Perspectives description.

Technology Perspective

Objective	Measure	Target
Partner with ELDIS for distribution network	Negotiate the contract with ELDIS	Contract by 01 December 2018
Producing green energy with Biogas	Biogas: total refurbishing of the existing 5 MWe plant and tuning with new turbine for increasing electrical and thermal efficiency by 5%	Date of achievement Project Plan
Producing green energy with pyrolysis from plastics	Purchasing new GT-Energy pyrolysis plant with a capacity of 2 MWe; installing, commissioning and operating	Developing the Business Plan If ROI $\leq$ 4 years then decision is positive
Energy Management and optimization via Blockchain contracts on Blockchain network	Build and own blockchain network that includes all customers of the heat and electricity network Add an AI module for evaluating trends and changes	Consult a specialist software company for Blockchain applications Project Plan

Objective	Measure	Target
	globally and per customer	
Heat distribution network by us COMPANY	Refurbish entire Heat network	Project Plan Date of achievement
Smart Metering	Collect Smart Metering readings and invoice via Blockchain network	Business Plan Project Plan for installing Smart Metering application

People Perspective

Objective	Measure	Target
Competent staff on energy generation with Biogas and Pyrolysis	Staff needs to be trained on Biomass and Pyrolysis recipe design and biogas process optimization	Work out staff training plan Execute staff training plan
Blockchain energy management know-how	Training to be provided on Blockchain building, design, and management concepts. It will be useful to hire a Blockchain specialist	Execute staff training plan
Blockchain contracts	Building know-how on developing	Execute staff training plan

Objective	Measure	Target
	contracts with customers via Blockchain technology and concept	Build internal Blockchain knowledge /expert team

Process Perspective

Objective	Measure	Target
Communication & Dissemination campaigns	Using public networks, website, advertisements on bills and regular publications, and ads in public media for awareness-building about green energy products and services. Using data from Smart Metering helps focus the messages towards the main consumption patterns	Outsource communication and dissemination to external expertise
E-Management	Implementing a global Energy Management is a critical element for all energy generation projects. E-Management on Blockchain will enable the	Evaluate ISO50001 combined with Blockchain data collection and AI decision-making. Use external ISO and Blockchain expertise

Objective	Measure	Target
	company to obtain real-time consumption measures from the entire green network	

Process perspective (cont.)

Objective	Measure	Target
Optimization of production	Data obtained from the Blockchain consumer network and from the E-Management tool combined with Artificial Intelligence will best work for optimizing the energy production	Design Blockchain network Design E-Mgmt tool Design AI engine Obtain agreement from all customers on data collection Implement software tools and network
Proactive and AI-supported maintenance plan	Preventive maintenance plan assisted by AI will minimize the risk of network failure and guarantee maximum availability of the network	Consult expert on CMM (computerized maintenance management) Contract CMM tool Implement CMM tool

Production Perspective

Objective	Measure	Target
Maximizing efficiency	Make the energy generation energy efficient by optimizing generation processes and full reuse of waste thermal energy	Analyze energy generation system for process optimization and energy efficiency. Use Pinch analysis for waste heat recovery
Production capacity usage	The energy production capacity should always be aligned to the consumption requirements. Higher production capacity versus lower consumption produces waste energy and makes the system less efficient	Design energy control system for on-demand usage and production. Implement ECS

Customer Perspective

Objective	Measure	Target
One-stop shopping for heat and electricity	Selling the customer the energy bundle keeps your revenue up and costs low. Customer satisfaction is high because the customer has only one contact point and one contract for both energy supplies	Develop energy bundle contracts. Test contracts with existing customers. Align and implement
Affordable renewable heat and electricity and providing energy supply contracts with discount for buying both heat and electricity	Keep the energy tariffs affordable by reducing the generation costs and increasing the generation efficiency with dual contracts (electricity and heat)	Measuring the energy generation and costs by permanently comparing with sales revenues
Residential customers	Limiting your offer to residential customers means keeping the costs under control by using only local/residential networks	Define the boundaries of a residential offer. Offer limited to residential customers

Emissions/Environmental Perspective

Objective	Measure	Target
Primary energy emission neutral	By using Biomass as primary energy substance, we keep the primary energy generation emission neutral	Emissions at level zero
Residual Biomass waste reduction	Process the residual biomass containing effluents from pig and/or cow manure and agricultural residues for separating liquids and solids. Processed liquids don't contain nitrates and will serve for agricultural irrigation. Solids will be processed to fertilizer pellets	Consult expertise for liquid and solid conversion Evaluate sales costs for solid fertilizer to farmers Produce a marketing plan Produce a business plan Project Plan for process implementation
Emission changes by process optimization	Optimizing processes will contribute to increase of efficiency and thus lower secondary emissions	Measure best performance by E-management system
CO_2 certificates collected in period	Green certificates will contribute to increase finance performance	Contract with trading company

Impact Perspective

Objective	Measure	Target
Impact on energy savings measures on production process	Energy savings measures will have an immediate (direct) impact to production process optimization	Produce a permanent energy efficiency measures program to be monitored by the E-management system
Impact on energy savings measures proposed to clients	A fixed part of the energy savings can be proposed in promotional actions as savings for clients	Evaluate regular promotional actions for sharing part of the savings
New customers	New customers help increase the overall performance of the generation system, by using the available free generation capacity	Implement special bonus for sales staff when contracting a new customer
Leaving customers	Leaving customers have a negative effect on finance and production performance	Avoid customers leaving by encouraging sales staff on farming existing customers Give bonus to sales staff for loyal customers with 2, 5, 10 years and more with the company Provide loyal customers with anniversary bonus

Economic Savings Perspective

Objective	Measure	Target
Savings by change on impacts	Costs are directly affected by increasing energy efficiency measures and thus generating direct savings on energy-related activities	Implement staff bonus system for providing effective measures on energy efficiency
Internal costs savings	ISO 50001 energy management system will help maximize energy efficiency, staff competence, and customer satisfaction	Make everybody in the company align to the E-Management system and to provide maximum customer satisfaction. Develop permanent customer satisfaction surveys

Development of the Key Performance Indicators

Once the eight perspectives have been completed with objectives, measures, and targets, we must extract the measuring elements from the target descriptions. These measurement elements are also named Key Performance Indicators; we measure the status of execution of a target action and Key because these indicators are critically (Key) relevant for monitoring the process.

On Table 5 we see the KPI table with all the attributes that are required for controlling the KPI, and for planning corrective actions in case the KPI is drifting above or below its set point.

Objective	Measure	Target	Requirements	KPI Id (Key Performance Indicator)	Unit	Target value	Data Period	Data sources	Budget EUR annual	Owner
Partnership with EDISON distribution network	Be part of the EDIS network	Contract by December 2018	technical know-how, Legal knowledge	TE001	Date per period	Date per period	Q4 2018			Peter
Producing green energy with Biogas	Biogas installation for the absorption of 5 MW plant and turning with a turbine for increasing the electrical and thermal efficiency by 5%	Part achievement of the Project Plan	Engineering, Construction, Purchasing, Logistics	TE002	Date per period	Date per period	Q1 2019	DB Engineering	€ 5 Mio	John
Producing green energy with Pyrolysis from plastics	Purchasing new 6 MW GE-Energy pyrolysis plant with a capacity of 2 MW We, installing, commissioning and operating	Developing the Business Plan for ROCH 4 years and the decision for positive	Engineering, Construction, Purchasing, Logistics	TE003	Date per period	Date per period	Q2 2019	DB Engineering	€ 6 Mio	name
Energy Management via Blockchain optimization on Blockchain	Build own blockchain network that includes all customers of the customer sold Add-Modul for evaluating trends and charging logs by a per customer	Count as a paid specialist software company for Blockchain applications for Project Plan	Software Engineering, Blockchain, Purchasing	TE004	Date per period	Date per period	Q3 2019	DB Engineering	€ 2 Mio	name
Blockchain network	Management of customers on the blockchain		Blockchain management	TE005	number of customer contracts	10.000 1st year +- 5%	2020	BC-DB 001	€ 160.000	name
	Consumer trend measurements			B006	MWh	within limit of 15% +-	2020	BC-DB 001	€ 60.000	name

Table 5: Development of the KPIs [14]

[14] Source: Georges Seil, BIZ-Consultant sarl

As for Fig.7, all KPIs related to these eight perspectives need to be developed and extended. We limited the development to the Technology Perspective for demonstration purposes.

The first three columns with Objective, Measure, and Target are the descriptions copied from our prior perspective analyses. The requirements indicate what kind of special requirements we need for executing the KPI monitoring and the perspective action. Each of the KPIs has a name. In our case, we named them according to the Perspective "Technology" with TE and an increasing reference number. All other KPI names are valid if one can find a direct relation with the perspective.

The KPI needs a unit for measurement. In our case, we have dates, periods, number of contracts, and MWhe. The target value indicates the value we want to achieve. In some cases, as for our example, a target value can be the same as the date of achievement. The data source is important to know when using data from and to the energy management system or a project management plan. In most cases, a data source will have to be interfaced with another data type, which will result in the design and build of a data interface between two heterogenous systems. The annual budget determines how much money is required for executing the action/task/objective. An additional KPI is sometimes required for monitoring the budget consumption when executing the action plan. Each KPI must have an owner who is responsible for the management of his KPI(s).

Beside the KPI table, the next step will be the development of a risk analysis with a risk mitigation plan for each of the KPIs. This is important in case the KPI's real value drifts away from the target value, in order to know the causes and the measures for taking a corrective action.

The entire Balanced Scorecard development is not the subject of this book. The objective of this book, based on Doctoral research, is limited to demonstrating all strategic elements that are required for developing and managing an effective powerful energy strategy.

Conclusion

The book analyzes different types of strategies for the use on energy projects and concludes that Porter's competitive strategies are only valid in specific circumstances where competition is expected. In our case, regarding the EU Program projects, Blue Ocean strategy is the best way for encouraging highly innovative thinking that allows differentiation from the crowd.

In this case, it would be most useful to consider both types of strategies at the beginning of the strategy development process, thus being able to test competitive and innovative measures and tactics, prior to deciding which way to go.

The Balanced Scorecard remains one of the best tools when combined with the Business Model Canvas (BMC) philosophy. Business Model Canvas makes us think about what the optimal business model could be for a specific case of an energy project. Based on the outcome of the BMC, the development of strategic themes and the strategy maps, including the development of the objectives and KPIs, will be the business of the Balanced Scorecard. The combination of tools builds the strength of this type of analysis.

Blockchain technology is a new technology that is currently mainly focused on financial transactions and virtual money, but its future will be oriented towards economic and industrial applications, such as those listed in the last section. In a next book we will dig deeper into the concepts of Blockchain and compare the pros and cons between Blockchain and the Ethereum network. Book II- "Energy and Blockchains" will provide some good examples on how to build a Blockchain or Ethereum network, with some programming examples on specific functional blocks.

Bibliography and References

https://ec.europa.eu/energy/en/topics/energy-strategy-and-energy-union/building-energy-union

http://www.balancedscorecard.org/BSC-Basics/About-the-Balanced-Scorecard

Robert M. Grant; Contemporary Strategy Analysis; 2002; Blackwell Publishers

https://ec.europa.eu/commission/priorities/energy-union-and-climate/fully-integrated-internal-energy-market_en

Table 2: Energy Model Canvas developed by the author from the Source: Alexander Osterwalder; 2008; Business Model Generation. www.businessmodelgeneration.com

Crosbie, Tracey & Short, Michael & Dawood, Muneeb & Shvadron, Uzi & Ala-Juusela, Mia & Rosqvist, Johanna & Brassier, Pascale & Stahlberg, Maarit & Thibault, Emilie & Gras, Denis & Project, Ideas. (2014). Specific Business Models for Demo Cases (D2:2).

Fig 3: M. Porter; Generic Strategies
[Source: TUTOR2U – study notes
https://www.tutor2u.net/business/reference/porters-generic-strategies-for-competitive-advantage]

https://www.blueoceanstrategy.com/about-the-authors/
[Source: Renée Mauborgne and W. Chan Kim; 2015; Harvard Business Review Press]
W. Chan Kim and Renée Mauborgne; Blue Ocean Strategy; Harvard Business School Press 2005

[The Balanced Scorecard; Robert S. Kaplan and David P. Norton; 1996]
Other sources: Slideshare by LinkedIn; Georges Seil; BIZ-Consultant sàrl

Author's short Biography

Georges Seil, PhD was born 1952 in Luxembourg as a Luxembourg citizen.

Georges studied electrical engineering and computer science in Luxembourg and Germany and started his professional career in the steel industry. In early 2000's, he specialized in Strategic Energy Management for complex industrial systems and became a Certified Energy Manager in 2013 (C.E.M.®), accredited ESOS Lead Auditor, ISO 50001 Auditor in 2014 and Certified Measurement and Verification Professional in 2019 (C.M.V.P.®).

In August 2019, Georges graduated as a Doctor of Philosophy in Energy Strategy.

He accumulated his extensive site and project experience in Western Europe, all Balkan countries, the Ukraine, Turkey, and the Maghreb countries.

Since 2006 he has worked as a registered Senior Advisor and Energy Specialist for the European Bank for Reconstruction and Development. Since 2015 Georges became official project evaluator as Energy Expert by the European Commission. Dr. Seil started his own Management Consulting Company in 2002, advising companies on performance improvement and strategic development, and advising industrial and service companies and the public sector on applying the Balanced Scorecard methodology. Georges adapted the Balanced Scorecard as a strategy design and management tool to be applied in the domain of Energy Efficiency and EE Finite Elements analysis.

With his electrical engineering background and more than 40 years of Management experience, Georges Seil gained substantial experience and know-how in the industrial sectors of metal, pulp and paper, printing, food production, wood processing, retail, engineering, construction, telecom, ITC, plastic recycling, WEE recycling, and pyrolysis systems.

His company BIZ-Consultant sàrl is incorporated in the Grand Duchy of Luxembourg, his country of birth and of residence.

Dr. Seil is a lecturer at the Polytechnic University of Timisoara and offers several energy- and strategy-related trainings by his own company in Luxembourg and in all his client countries. His most recent provided training included ISO 50001 introduction and implementation for industries in Casablanca, Morocco, in cooperation with AOB Group Casablanca.

Dr. Georges Seil is a Member of the Association of Energy Engineers (AEE) and Senior Member of the Institute of Electric and Electronic Engineers (IEEE).

This book is his second publication of a planned series of energy-related books.

New publications will be announced on LinkedIn and his website www.biz-consultant.net.

Contact: georges.seil@biz-consultant.net

LinkedIn profile: https://www.linkedin.com/in/georges-seil-phd-60027a?lipi=urn%3Ali%3Apage%3Ad_flagship3_profile_view_base_contact_details%3B2kVl9XZETxGhIC%2F2U5T5yQ%3D%3D